# LEARNING TIME WORKBOOK

FOR KIDS 6-10

Book Title: Learning Time Workbook

Published by "Light of Mind Publishing"

Author: Elisabeth Richard



ISBN: 978-1-7387571-7-6

# THANK YOU FOR TRUSTING US WITH YOUR CHILD'S EDUCATION

Light of Mind
PUBLISHING

EDUCATIONAL
ACTIVITY WORKBOOKS

## CHECK OUT OUR OTHER BOOKS:

https://linktr.ee/lightofmindhomeschool

Scan our QR Code

To contact the publisher

lightofmindpublishing@gmail.com

# This Book Belongs To:

# UNDERSTANDING TIME

= IN A DAY

**A day is divided into 24 hours.**

**Each hour having 60 minutes and each minute having 60 seconds.**

**This means there are 1440 minutes and 86,400 seconds in a day.**

= 1 HOUR

**You can think of the day like a circle that starts at 12 o'clock, and as the day goes on, the clock tells the time from morning (6:00 AM) to evening (6:00 PM), until it starts over again at midnight (12:00 AM).**

LOOK AT THE MINUTES AND THE HOURS

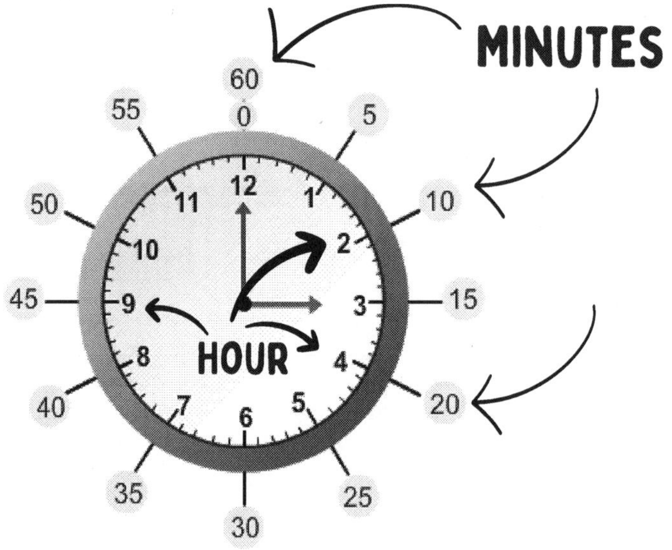

# LOOK VERY CLOSELY AT THE HOURS AND MINUTES

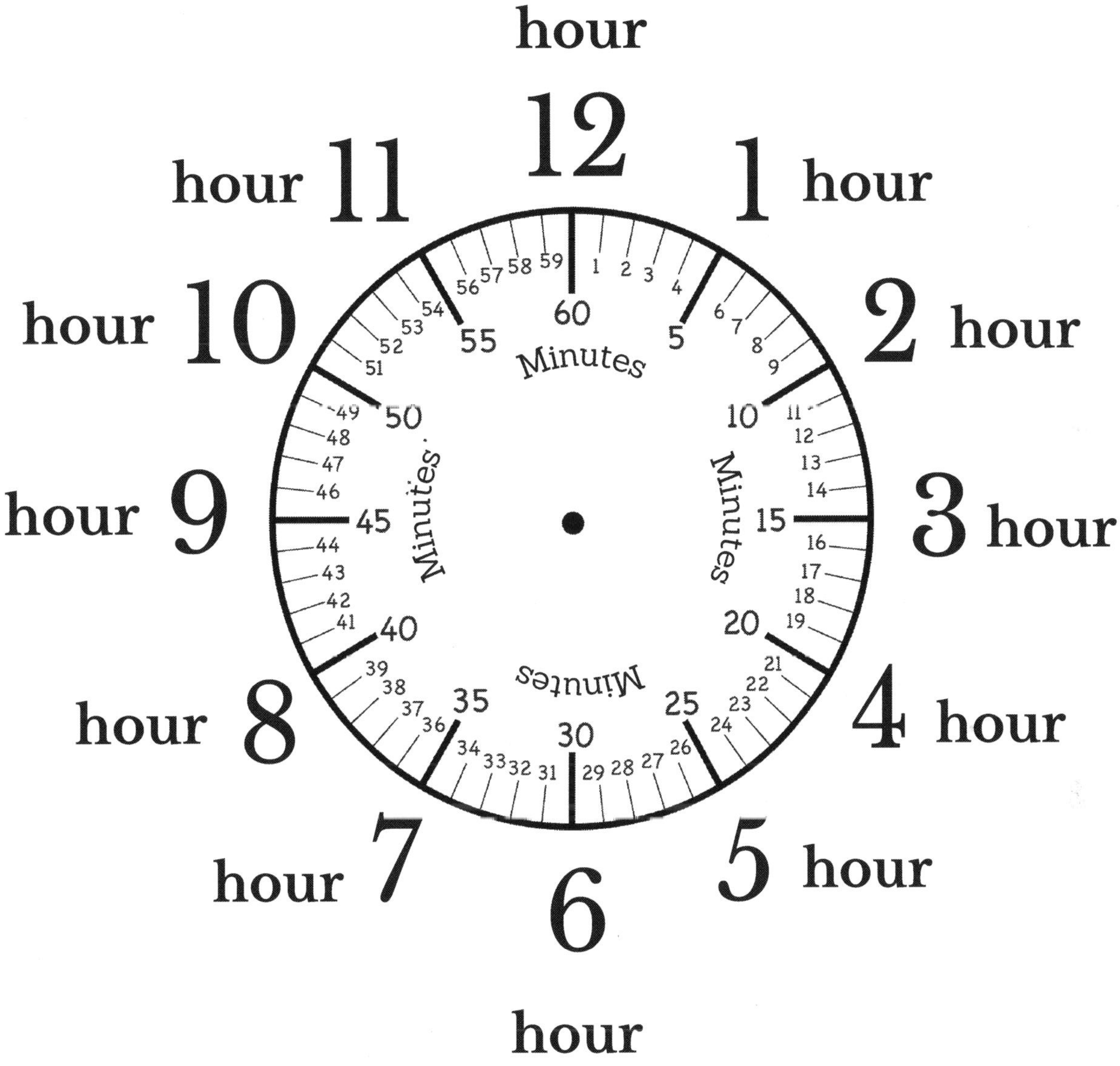

# IDENTIFY THE MINUTES ON THIS CLOCK

WRITE THE ANSWER ON THE LINE.

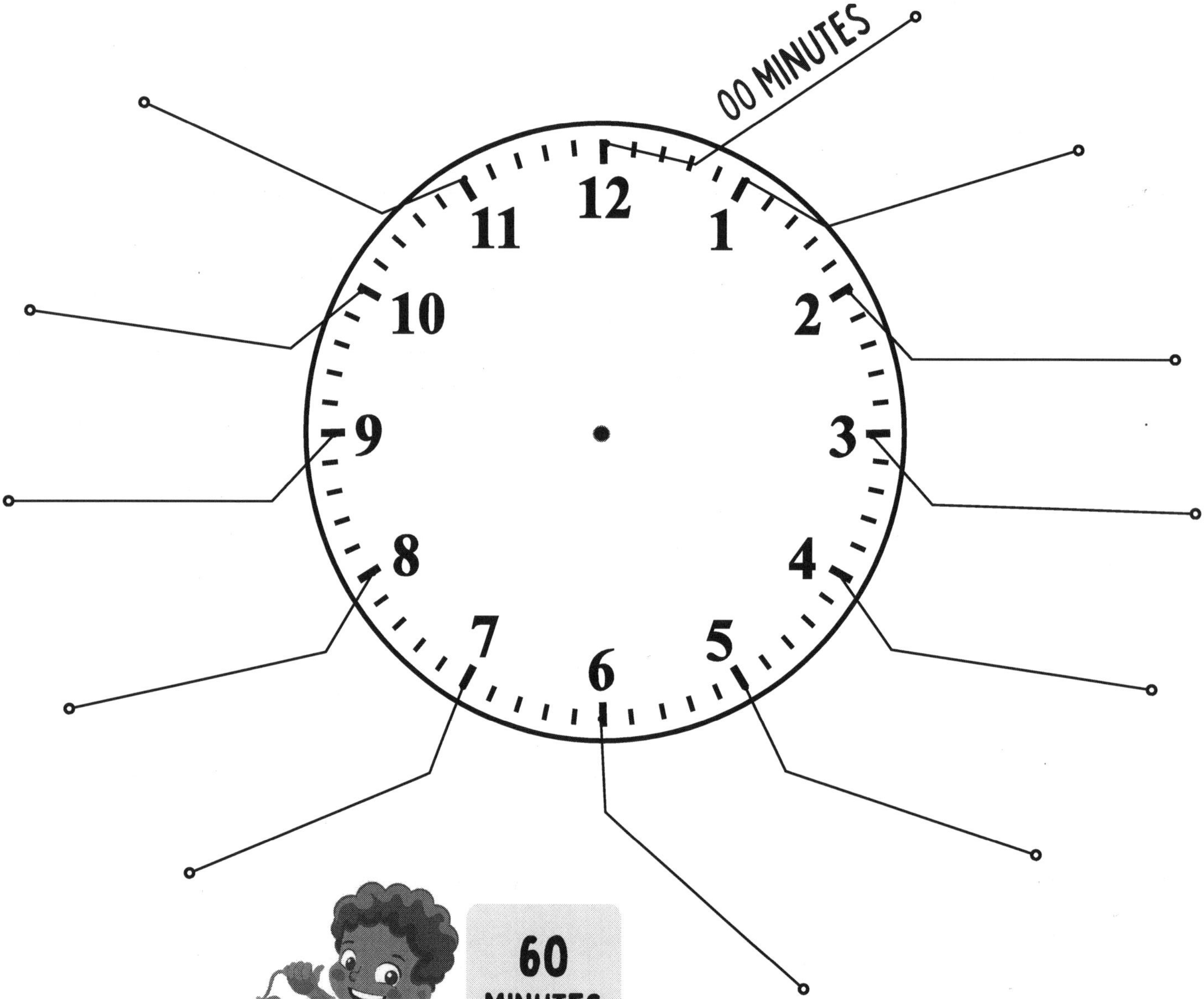

60 MINUTES = 1 HOUR

# IDENTIFY THE HOURS ON THIS CLOCK

WRITE THE ANSWER ON THE LINE.

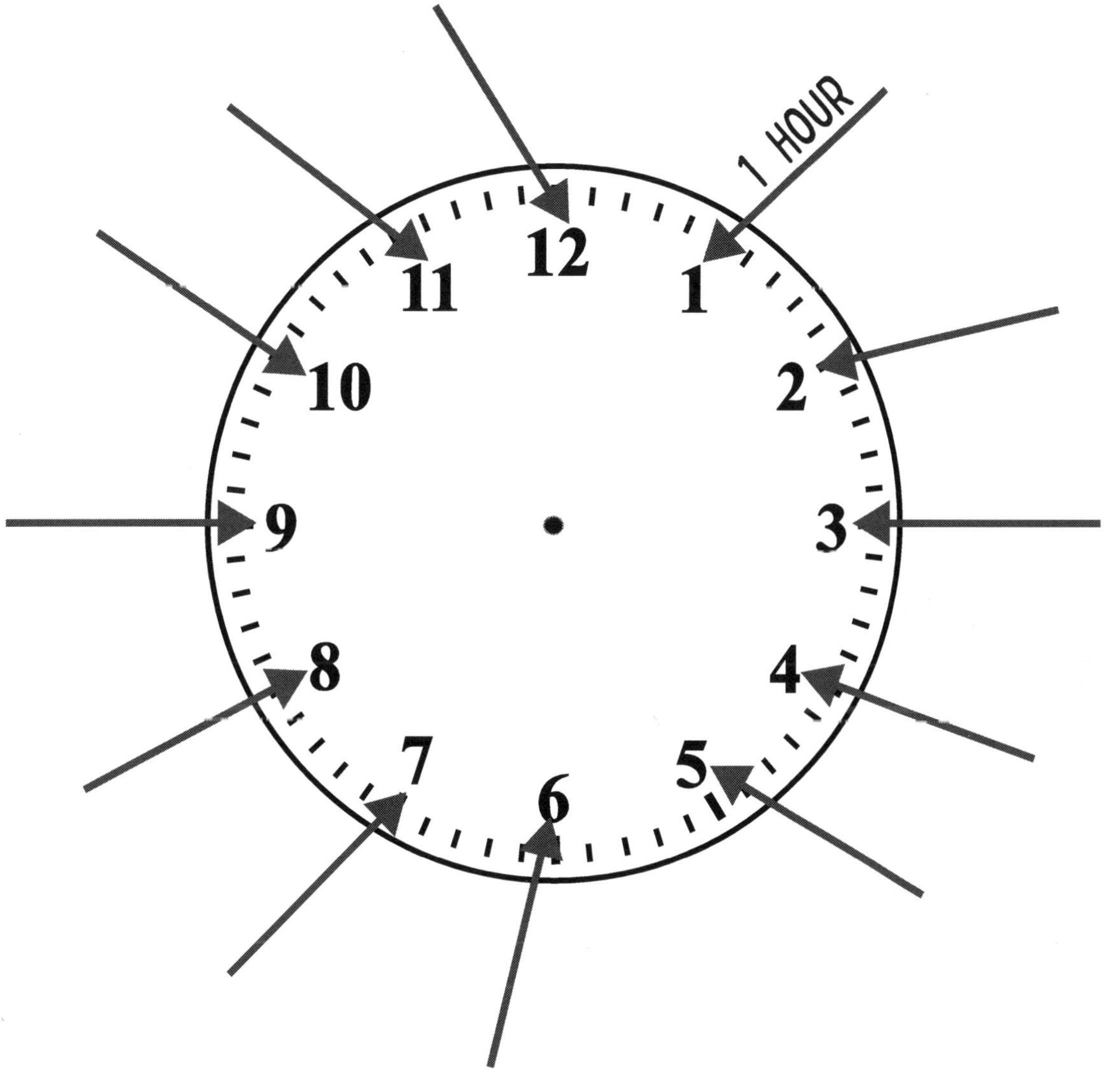

# STUDY THE HANDS OF THE CLOCK

WHAT NUMBERS ARE THE LONG HAND AND SHORT HAND POINTING TO?

LONG HAND = _________
MINUTES

SHORT HAND = _________
HOURS

# A.M. AND P.M.

Do you know what "a.m." and "p.m." mean?

They are special words we use to tell time.

When we say "a.m.," it means that it is morning time. The "a" in "a.m." stands for "ante," which is a Latin word that means "before." So, "a.m." means "before noon."

On the other hand, when we say "p.m.," it means that it is afternoon or evening time. The "p" in "p.m." stands for "post," which is another Latin word that means "after." So, "p.m." means "after noon."

Remember, when it is morning time, we say "a.m.,"
and when it is afternoon or evening time, we say "p.m."
It's a special way to tell time!

A.M.. is from 12 o'clock at night to 12 o'clock noon.
P.M. is from 12 o'clock noon to 12 o'clock at night.

MORNING A.M.

A.M.= MIDNIGHT TO 12:00 NOON

(BEFORE NOON)

AFTER 12:00 NOON, 1 P.M.-4:00 P.M.

AFTERNOON

5 P.M. 6:00 P.M.

7 P.M. EVENING

8 P.M. TO 12:00 MIDNIGHT

NIGHT

# A.M. AND P.M.

READ THE SENTENCES AND CIRCLE WHETHER THE EVENTS HAPPENED IN THE A.M. OR P.M.

JAMES IS GETTING UP TO GO TO SCHOOL.

A.M. OR P.M.

BOBBY IS PUTTING ON HIS PAJAMAS.

A.M. OR P.M.

THE CHILDREN AT SCHOOL ARE EATING LUNCH. IT IS 11:45.

A.M. OR P.M.

# A.M. AND P.M.

READ THE SENTENCES AND CIRCLE WHETHER THE EVENTS HAPPENED IN THE A.M. OR P.M.

PEDRO JUST CAME HOME FROM SCHOOL AND IS EATING FRUIT.

A.M. OR P.M.

HENRY IS EATING SPAGHETTI THIS EVENING.

A.M. OR P.M.

MARY ALWAYS FEEDS THE CHICKENS BEFORE SHE LEAVES FOR SCHOOL.

A.M. OR P.M.

# A.M. AND P.M.

READ THE SENTENCES AND CIRCLE WHETHER THE EVENTS HAPPENED IN THE A.M. OR P.M.

IT'S 12:30 AND GRACE IS ENJOYING HER BURGER AND FRENCH FRIES.

A.M. OR P.M.

KAYLA AND JORDAN JUST FINISHED DINNER. THEY ARE NOW BAKING COOKIES FOR DESSERT

A.M. OR P.M.

ELIZABETH HAS BALLET PRACTICE EVERY SATURDAY MORNING.

A.M. OR P.M.

# HELP THESE CHILDREN BE ON TIME

READ THE STORY. THEN CIRCLE THE TIME
THAT MATCHES THE CLOCK.

BRIAN STARTS SOCCER PRACTICE AT ?

5:00 6:00 4:00

MARTHA PLAYS TENNIS ON SATURDAYS AT?

9:00 8:00 10:00

# WHAT TIME IS IT?

READ THE STORY. THEN CIRCLE THE TIME
THAT MATCHES THE CLOCK.

JULIA ATE HER LUNCH AT?

12:30 11:30 12:00

CHARLES PLAYS BASKETBALL AFTER SCHOOL AT?

3:30 2:30 4:30

# HELP THESE CHILDREN BE ON TIME

READ THE STORY. THEN CIRCLE THE TIME THAT MATCHES THE CLOCK.

KATIE PRACTICES HER KEYBOARD EVERY DAY AT?

5:30 7:30 3:30

EMMANUEL IS GOING TO HIS FRIEND'S BIRTHDAY PARTY AT?

11:30 2:30 1:30

BETHANY GOES TO BED AT?

7:30 9:30 6:30

# HELP THESE CHILDREN BE ON TIME

READ THE STORY. THEN CIRCLE THE TIME
THAT MATCHES THE CLOCK.

MANDY WASHED HER TEETH AT ?

9:30 8:30 10:15

MARCO IS STARTING HIS HOMEWORK.
WHAT TIME IS IT?

3:35 4:30 2:30

BOBBY HAS A SPELLING TEST AT?

9:30 10:30 3:30

# HELP THESE CHILDREN BE ON TIME

READ THE STORY. THEN CIRCLE THE TIME
THAT MATCHES THE CLOCK.

MY FRIEND IS COMING OVER AT?

7:30 7:25 7:15

GRANNY STARTS KNITTING AT?

3:40 3:30 3:25

GRANNY NEEDS TO DELIVER
HER PACKAGES AT?

9:15 10:10 9:10

# READ THE TIME AND DRAW THE HANDS ON THE CLOCK

2:30

10:00

5:15

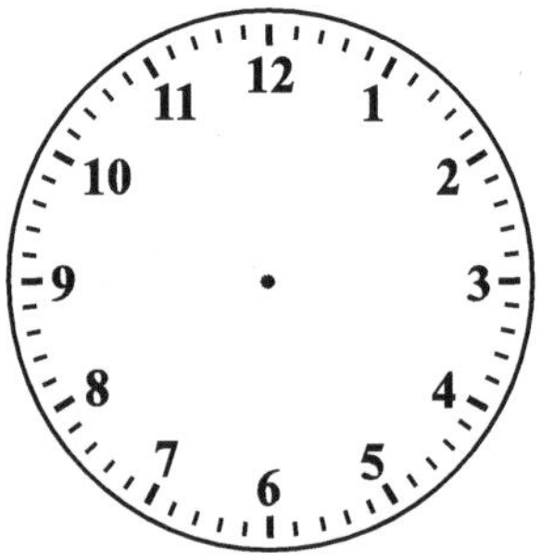

11:35

4:15

7:25

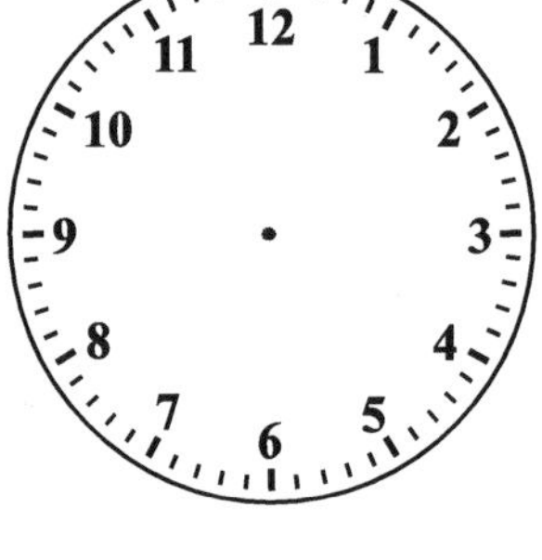

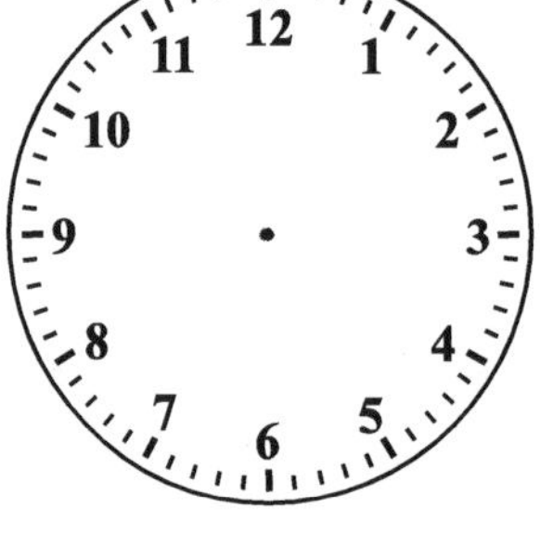

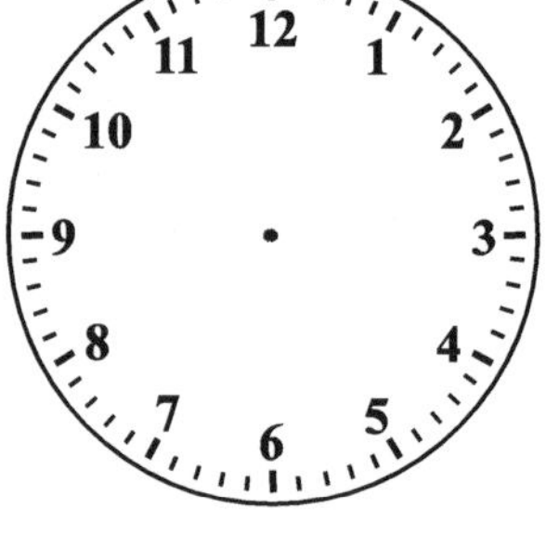

3:50

9:10

8:40

# READ THE TIME AND DRAW THE HANDS ON THE CLOCK

6:10

8:05

11:55

1:45

2:15

3:35

4:20

9:20

7:30

# READ THE TIME AND DRAW THE HANDS ON THE CLOCK

12:20

5:35

7:00

12:25

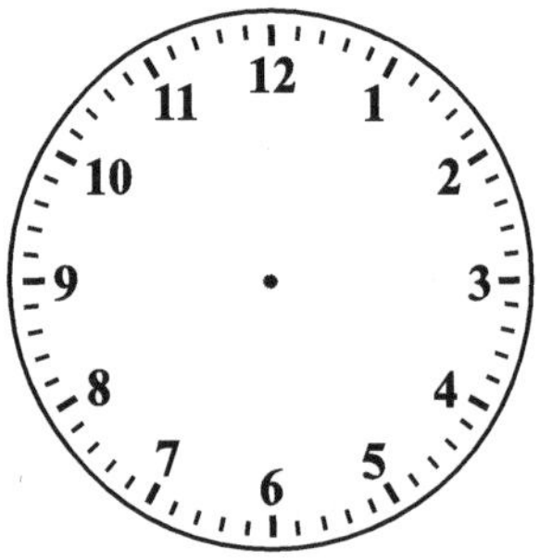

4:40

3:25

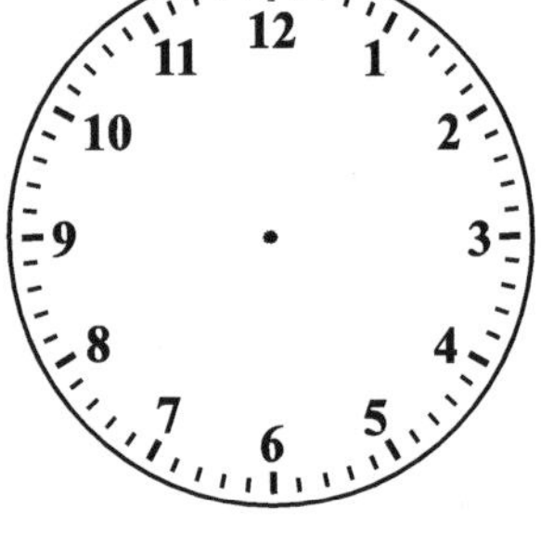

2:35

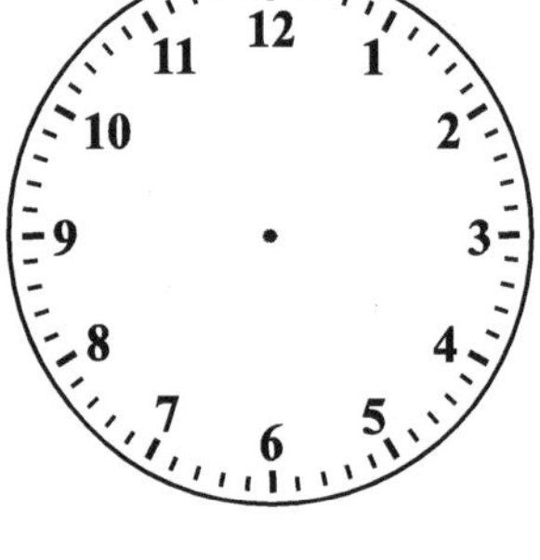

1:55

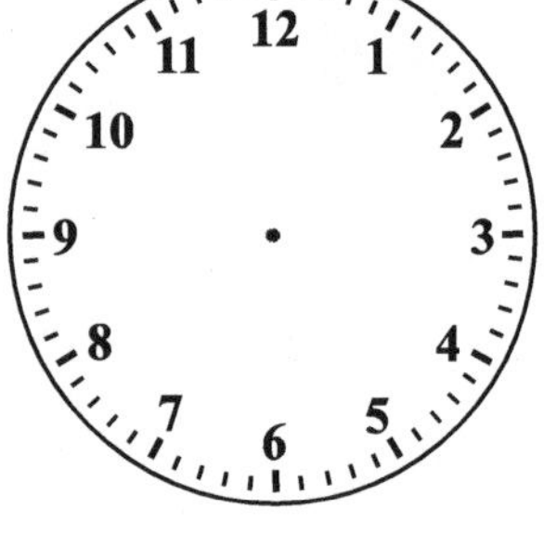

8:10

# READ THE TIME AND DRAW THE HANDS ON THE CLOCK

3:15

4:10

9:30

10:10

1:00

6:30

9:45

11:05

5:50

# WRITE THE CORRECT TIME

3:00

# WRITE THE CORRECT TIME

# WRITE THE CORRECT TIME

# WRITE THE CORRECT TIME

# HOW MUCH TIME DOES IT TAKE YOU TO BRUSH YOUR TEETH?

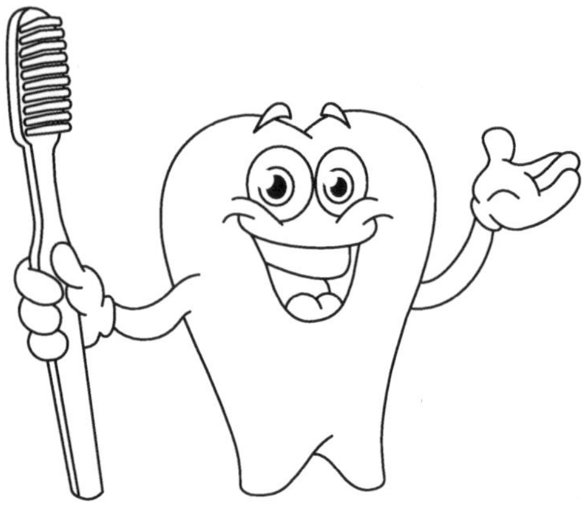

To properly wash our teeth and get rid of all the plaque, we need to brush them for a total of two minutes. That's just like singing your favorite song or counting from 1 to 120!

2 minutes

Color me!

# TIME
# WORD PROBLEMS

**Problem 1:**

Tommy's family went to the zoo to see the lions. They arrived at the zoo at 10:00 a.m., and spent 60 minutes watching the lions. After that, they went to see the monkeys for 30 minutes.

What time did Tommy and his family finish seeing the monkeys?

Use this space to do your calculations:

# TIME
# WORD PROBLEMS

**Problem 2:**

Samantha wants to bake a cake for her friend's birthday party, which starts at 3 o'clock in the afternoon.

It takes her 1 hour and 30 minutes to bake the cake.
What time does Samantha need to start baking the cake so that it will be ready in time for the party?

Use this space to do your calculations:

# TIME
# WORD PROBLEMS

**Problem 3:**

Lucas woke up at 6:30 am to get ready for school.
It takes him 20 minutes to eat breakfast, 5 minutes to brush his teeth, and 10 minutes to get dressed.

What time will Lucas need to leave for school?

Use this space to do your calculations:

# TIME
# WORD PROBLEMS

**Problem 4:**

Emma started reading a book at 4:15 pm. She read for 35 minutes and then took a 15-minute break to have a snack.

After her break, she read for another 40 minutes to finish the book.

Use this space to do your calculations:

# TIME
# WORD PROBLEMS

**Problem 5:**

Mrs.Johnson started teaching a math class at 9:15 am.
She taught for 45 minutes and then took a 15-minute break for lunch.

After her break, she taught for another 45 minutes.

What time did Mrs. Johnson finish teaching her math class?

Use this space to do your calculations:

# TIME
# WORD PROBLEMS

## Problem 6:

Andrew started feeding the chickens on his family's farm on Monday morning at 6:00 am.

It takes him 10 minutes to fill up the chicken feeders and 5 minutes to collect the eggs.

After he's finished with the chickens, he goes to milk the cows. Milking each cow takes 15 minutes, and there are 2 cows to milk.

What time will Andrew finish his chores on the farm?

Use this space to do your calculations:

Use this space to do your calculations:

# TIME
# WORD PROBLEMS

## Problem 7:

Olivia is an 8-year-old girl who loves playing the piano after school. She starts practicing at 3:30 PM and practices for 45 minutes every day.

What time will Olivia finish practicing the piano?

Use this space to do your calculations:

# TIME
# WORD PROBLEMS

## Problem 8:

Jackson is a 9-year-old boy who loves playing hockey on Saturdays. He usually plays for 1 hour and 15 minutes.

If he finishes playing at 11:45 AM, what time did he start playing?

Use this space to do your calculations:

# TIME
# WORD PROBLEMS

## Problem 9:

One sunny afternoon, a boy named Kevin and his dog, Rocky, decided to go for a run.

They started at 3:00 PM and ran for 30 minutes before stopping to drink water for 10 minutes.

Kevin was thirsty and drank water for the entire duration. At what time did Kevin and Rocky resume their run?

Use this space to do your calculations:

# TIME
# WORD PROBLEMS

## Problem 10:

One day, a girl named Annabelle was playing with her cat, Whiskers, in the park. They arrived at 10:00 AM and played for 1 hour.

Then, Annabelle stopped to buy an ice cream from an ice cream truck in the park, and it took her 15 minutes to get the ice cream and pay for it. At what time did they leave the park?

Do you remember how many minutes = 1 hour?

Use this space to do your calculations:

# COLOR ME !

# TIME

## Common English Vocabulary We Use To Tell The Time

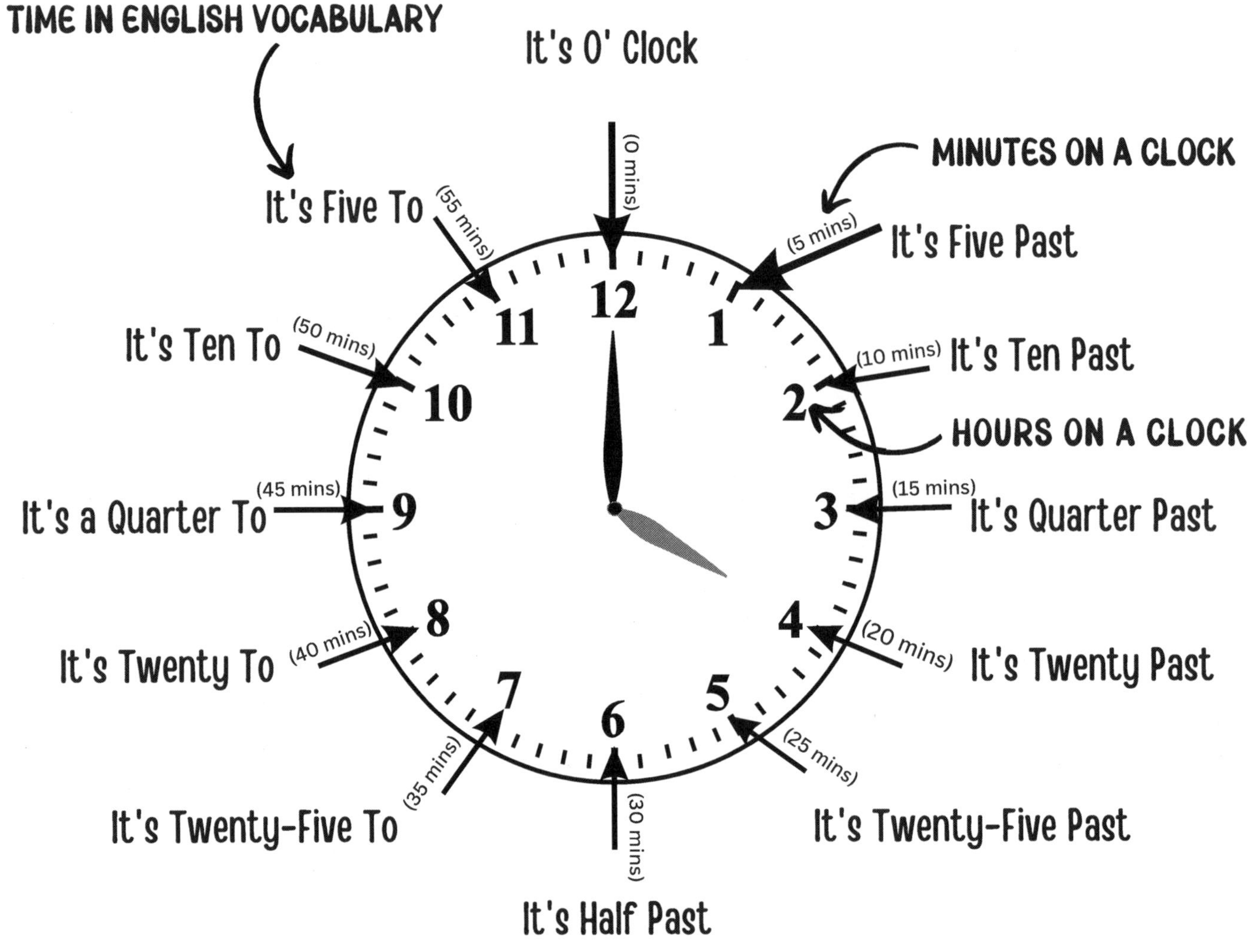

# MATCH THE TIME

Draw a line from the time worded in English
to the appropriate clock.

QUARTER TO 4

TEN PAST 10

QUARTER PAST 6

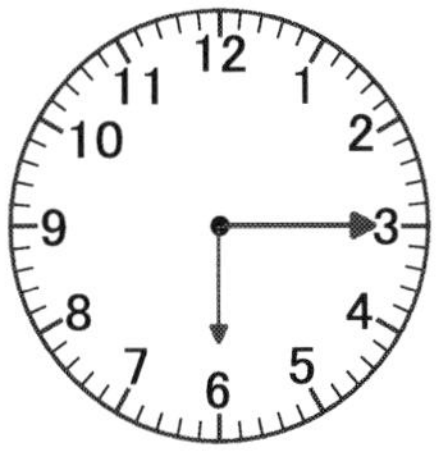

HALF PAST 7

TWENTY TO 9

# MATCH THE TIME

Draw a line from the time worded in English
to the appropriate clock.

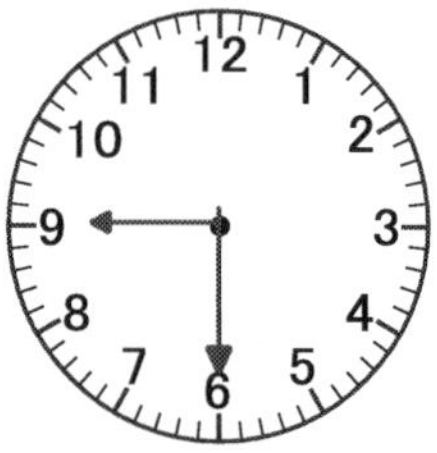

QUARTER PAST 11

TEN PAST SEVEN

THREE O' CLOCK

HALF PAST NINE

TWENTY-FIVE TO TWO

TWENTY-FIVE PAST ONE

# MATCH THE TIME

Draw lines from the time said in English to the correct time in numbers on a clock.

| | |
|---|---|
| 1. One O'Clock | A. 1:15 |
| 2. Half Past 1 | B. 1:45 |
| 3. A Quarter To 2 | C. 1:30 |
| 4. A Quarter Past 1 | D. 1:00 |
| 5. Twenty Past 1 | E. 1:25 |
| 6. Ten Past 1 | F. 1:50 |
| 7. Twenty-Five Past 1 | G. 1:35 |
| 8. Five To 2 | H. 1:40 |
| 9. Ten To 2 | I. 1:10 |
| 10. Twenty To 2 | J. 1:55 |
| 11. Twenty-Five To 2 | K. 1:20 |

## COLOR ME!

# MATCH THE TIME

Draw lines from the time said in English to the correct time in numbers on a clock.

| | |
|---|---|
| 12. Three O'Clock | A. 4:15 |
| 13. Half Past 6 | B. 2:45 |
| 14. A Quarter To 3 | C. 6:30 |
| 15. A Quarter Past 4 | D. 3:00 |
| 16. Twenty Past 4 | E. 12:25 |
| 17. Ten Past 5 | F. 12:50 |
| 18. Twenty-Five Past 12 | G. 11:35 |
| 19. Five To 4 | H. 10:40 |
| 20. Ten To 1 | I. 5:10 |
| 21. Twenty To 11 | J. 3:55 |
| 22. Twenty-Five To 12 | K. 4:20 |

## COLOR ME!

# ELAPSED TIME EXERCICES

| The Present Time | 5 Minutes Before | 10 Minutes After |
|---|---|---|
| 2:50 | 2:45 | 3:00 |
| 4:00 | | |
| 5:15 | | |
| 1:30 | | |
| 2:15 | | |
| 5:45 | | |
| 6:10 | | |
| 7:50 | | |
| 7:20 | | |
| 6:35 | | |

# ELAPSED TIME EXERCICES

| The Present Time | 5 Minutes Before | 10 Minutes After |
| --- | --- | --- |
| 3:40 | 3:35 | 3:50 |
| 2:00 | | |
| 9:15 | | |
| 12:35 | | |
| 10:15 | | |
| 3:10 | | |
| 5:25 | | |
| 11:00 | | |
| 1:20 | | |
| 8:30 | | |

# HOW TIME AFFECTS OUR LIVES

- 60 minutes make up one hour. Most people are awake for about 16 hours a day, so that's 16 hours x 60 minutes = 960 minutes.

- 24 hours = one day. That's why we have 24 hours in a day.

- Most people sleep for about 8 hours a day, so that leaves 16 hours for other activities.

- 7 days make up one week. That's why there are 7 days in a week. Sunday, Monday, Tuesday, Wednesday, Thursday, Friday, and Saturday.

- 4 weeks (or 28 days) = one month on the calendar. However, some months have 30 or 31 days instead.

- We use a calendar to keep track of which month it is and when special events are happening.

- 12 months make up one year. We start the year on January 1st and end it on December 31st.

# DAYS OF THE WEEK

- Sun
- Mon
- Tue
- Wed
- Thu
- Fri
- Sat

# MONTHS OF THE YEAR

1. JANUARY
2. FEBRUARY
3. MARCH
4. APRIL
5. MAY
6. JUNE
7. JULY
8. AUGUST
9. SEPTEMBER
10. OCTOBER
11. NOVEMBER
12. DECEMBER

Color the foods you eat on the different days of the week!

# HOW TIME AFFECTS OUR LIVES

## WHAT IS A LEAP YEAR?

There are **12 months** in a calendar year, and they are:

- January, February, March, April, May, June, July, August, September, October, November, December.

- Each month has a different number of days. January, March, May, July, August, October, and December have **31 days.**

- **February has 28 days**, except for **leap years** when it has **29 days**. April, June, September, and November have 30 days.

- A leap year is a special year that happens once every four years. **In a leap year, February has 29 days instead of 28 days.**

  This is because the Earth takes 365.25 days to orbit the sun, so an extra day is added to the calendar every four years to make up for the extra quarter day.

- Leap years are named by their year number. For example, the year 2020 was a leap year, and the next leap year will be 2024.

That's the basic breakdown of months in a calendar year and leap years!

# QUIZ TIME

1. How many hours are in a day?

   A) 12 hours
   B) 24 hours
   C) 48 hours
   D) 60 hours

2. How many days are in a week?

   A) 3 days
   B) 5 days
   C) 7 days
   D) 10 days

3. How many months are in a year?

   A) 6 months
   B) 10 months
   C) 12 months
   D) 24 months

4. How many seconds are in a minute?

   A) 10 seconds B) 30 seconds C) 60 seconds D) 90 seconds

5. How many minutes are in an hour?

   A) 10 minutes B) 30 minutes C) 60 minutes D) 90 minutes

# QUIZ TIME

6. How many days are in February (in a non-leap year)?

A) 28 days B) 30 days C) 31 days D) 29 days

7. Name the days in a week (in order).

A) Monday, Tuesday, Wednesday, Thursday, Friday, Saturday, Sunday
B) Tuesday, Wednesday, Thursday, Friday, Saturday, Sunday, Monday
C) Wednesday, Thursday, Friday, Saturday, Sunday, Monday, Tuesday
D) Sunday, Monday, Tuesday, Wednesday, Thursday, Friday, Saturday

8. Which months have 31 days?

A) January, March, May, July, August, October, December
B) February, April, June, September, November
C) January, February, March, April, May, June
D) July, August, September, October, November, December

9. How many days does February have in a leap year?

A) 28 days B) 29 days C) 30 days D) 31 days

10. What is the next leap year after 2024?

A) 2025 B) 2026 C) 2027 D) 2028

# WORD PROBLEMS
# CALENDAR DAYS & MONTHS

1. SOPHIE HAS A 10-DAY SUMMER VACATION. IF HER VACATION STARTS ON MONDAY, WHAT DAY OF THE WEEK DOES SHE RETURN TO SCHOOL?

VACATION

2. PABLO WANTS TO SAVE UP HIS ALLOWANCE TO BUY A TOY THAT COSTS $16. IF HE SAVES $2 A DAY, HOW MANY DAYS WILL IT TAKE HIM TO SAVE ENOUGH MONEY?

3. STACEY HAS A DANCE PERFORMANCE IN 9 DAYS. SHE WANTS TO PRACTICE FOR AT LEAST 30 MINUTES EVERY DAY UNTIL THE PERFORMANCE. HOW MANY HOURS WILL SHE PRACTICE IN TOTAL?

# WORD PROBLEMS
# CALENDAR DAYS & MONTHS

4. IF THE SCHOOL YEAR IS 180 DAYS LONG, HOW MANY DAYS OF SUMMER VACATION DO STUDENTS GET IF THE SCHOOL YEAR ENDS ON JUNE 5TH AND STARTS BACK UP ON SEPTEMBER 3RD?

5. JOHN HAS BEEN SAVING MONEY FOR 6 MONTHS. IF HE STARTED SAVING IN JANUARY, IN WHICH MONTH WILL HE HAVE SAVED FOR A WHOLE YEAR?

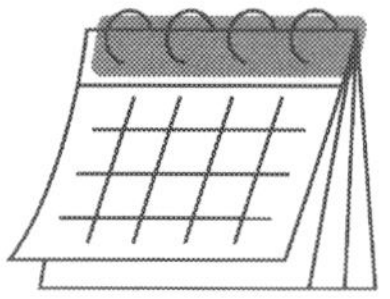

6. MARY'S BIRTHDAY IS IN 3 MONTHS. IF HER BIRTHDAY IS IN JUNE, IN WHICH MONTH IS IT NOW?

# ADVANCED QUIZ FOR KIDS
# GRADE 3-4

REMEMBER

60 MINUTES = 1 HOUR

1. Johnny has to do his chores every day after school. It takes him 15 minutes to make his bed, 25 minutes to clean his room, and 5 minutes to water the plants.

What is the total amount of time John spends doing his chores every day?

2. Sarah finishes her homework at 4:30 pm. If she spends 45 minutes watching TV, what time will it be when she is done?

A) 4:45 pm B) 5:00 pm C) 5:15 pm D) 5:30 pm

3. Katie is 9 years old and loves playing with her friends at the park. One day, they decided to have a friendly race from the swings to the slide. The distance between the swings and the slide is 50 meters. Katie can run 10 meters in 2 seconds.

Can you help Katie figure out how many seconds it will take her to reach the slide and win the race?

# ADVANCED QUIZ FOR KIDS
# GRADE 3-4

4. Joshua loves playing video games and wants to know how much time he spends gaming each week. He decides to keep track for 7 days and record his gaming time each day.

On Monday, Joshua plays games for 20 minutes. On Tuesday, he plays games for 30 minutes. On Wednesday, he has Cadet night and doesn't play any games, so his gaming time is zero minutes. On Thursday, he plays games for 35 minutes. On Friday, he plays games for 25 minutes.
On Saturday, he plays games for 25 minutes. On Sunday, he doesn't play any games, so his gaming time is zero minutes.

Can you help Joshua figure out the total time he spends gaming in a week?

Use the space below to do your calculations.

# ADVANCED QUIZ FOR KIDS
# GRADE 3-4

5. Cedric lives on a farm and takes care of the animals. One day, he decides to time how long it takes each animal to complete a specific task.

Here are the results: The chickens take 5 minutes to lay their eggs. The cows take 15 minutes to be milked. The pigs take 20 minutes to eat their food.
The horses take 30 minutes to be groomed.

Now, the question is:  How long will it take him to complete his chores?

# ANSWERS TO TIME WORD PROBLEMS (P.25-26)

**Solution to Problem 1:**

To find out what time Tommy and his family finished seeing the monkeys, we need to add the time they spent watching the lions to the time they spent watching the monkeys, and then add that total time to their arrival time.

The time Tommy and his family spent watching the lions was 60 minutes, which is the same as 1 hour.

The time Tommy and his family spent watching the monkeys was 30 minutes, which is the same as 0.5 hours.

So, the total time Tommy spent at the zoo was 1 + 0.5 = 1.5 hours.
To find out what time Tommy and his family finished seeing the monkeys, we need to add 1.5 hours to their arrival time of 10:00 AM:

10:00 AM + 1.5 hours = 11:30 AM

Tommy and his family finished seeing the monkeys at <u>11:30 AM</u>.

**Solution to Problem 2:**

To find out what time Samantha needs to start baking the cake, we need to subtract the time it takes to bake the cake from the time of the party:

3:00 PM - 1 hour and 30 minutes = 1:30 PM

So, Samantha needs to start baking the cake at <u>1:30 PM</u> to have it ready in time for the party at 3:00 PM.

# ANSWERS TO TIME
# WORD PROBLEMS (P.27-28)

**Solution to Problem 3:**

To solve the problem, add up the time it takes for Lucas to complete each task:
20 minutes (breakfast) + 5 minutes (brush teeth) + 10 minutes (get dressed) = 35 minutes.

Then, you need to add that 35 minutes to the time Lucas woke up:
6:30 am + 35 minutes = 7:05 am

Therefore, Lucas will be ready to leave for school at <u>7:05 am.</u>

**Solution to Problem 4:**

To solve the problem, add up the time Emma spent reading and on her break:

35 minutes (reading) + 15 minutes (break) + 40 minutes (reading) = 90 minutes

Then, you need to add that 90 minutes to the time Emma started reading:

4:15 pm + 90 minutes = 5:45 pm

Therefore, Emma finished reading her book at <u>5:45 pm.</u>

# ANSWERS TO TIME WORD PROBLEMS (P.29-30)

**Solution to Problem 5:**

To solve the problem, add up the time Mrs. Johnson spent teaching to the time she spent on her break:

45 minutes (teaching) + 15 minutes (break) + 45 minutes (teaching) =
1 hour and 45 minutes.

Then, you need to add that 1 hour and 45 minutes to the time Ms. Johnson started teaching:

9:15 am + 1 hour (60 minutes) and 45 minutes = 11:00 am

Therefore, Ms. Johnson finished teaching her math class at <u>11:00 am</u>.

**Solution to Problem 6:**

To solve the problem, add up the time Andrew spent feeding the chickens and collecting eggs:

10 minutes (feeding the chickens) + 5 minutes (collecting eggs) = 15 minutes.

Then, you need to add that 15 minutes to the time Andrew started his chores:
6:00 am + 15 minutes = 6:15 am.

Next, they can calculate how long it takes Andrew to milk both cows:

2 cows * 15 minutes per cow = 30 minutes.

Finally, add the milking time to the time Andrew finished with the chickens:
6:15 am + 30 minutes = 6:45 am.

Therefore, Andrew will finish his chores on the farm at 6:45 am on Monday morning.

# ANSWERS TO TIME WORD PROBLEMS (P.32-33)

**Solution to Problem 7:**

What time will Olivia finish practicing the piano?

Olivia practices for 45 minutes, so to find out what time she will finish practicing, we need to add 45 minutes to her start time of 3:30 PM.

To do this, we can add 30 minutes to 3:30 PM to get 4:00 PM, and then add another 15 minutes to get 4:15 PM.

Therefore, Olivia will finish practicing the piano at **4:15 PM.**

**Solution to Problem 8:**

To find out what time Jackson started playing, we need to subtract 1 hour and 15 minutes from his finishing time of 11:45 AM.

We then subtract 1 hour and 15 minutes from 11:45:
11:45 - 1:15 = 10:30

Therefore, Jackson started playing at **10:30 AM.**

# ANSWERS TO TIME WORD PROBLEMS (P.34-35)

**Solution to Problem 9:**

Kevin and Rocky started running at 3:00 PM and ran for 30 minutes until 3:30 PM.

They then stopped to drink water for 10 minutes, which means they resumed their run 10 minutes after 3:30 PM.

To determine the exact time, we can add the 10-minute water break to the current time, which is 3:30 PM.

Therefore, Kevin and Rocky resumed their run at **<u>3:40 PM.</u>**

**Solution to Problem 10:**

To find out what time Annabelle and Whiskers left the park, we need to add the time they spent playing and the time spent getting ice cream.

Annabelle played with her cat for 1 hour, so if they arrived at 10:00 AM, then they finished playing at 11:00 AM.

Next, Annabelle spent 15 minutes getting ice cream, so we need to add 15 minutes to 11:00 AM.

11:00 AM + 15 minutes = **<u>11:15 AM</u>**

Therefore, Annabelle and Whiskers left the park at 11:15 AM.

# ANSWERS TO P.38:

## MATCH THE TIME

Draw a line from the time worded in English
to the appropriate clock.

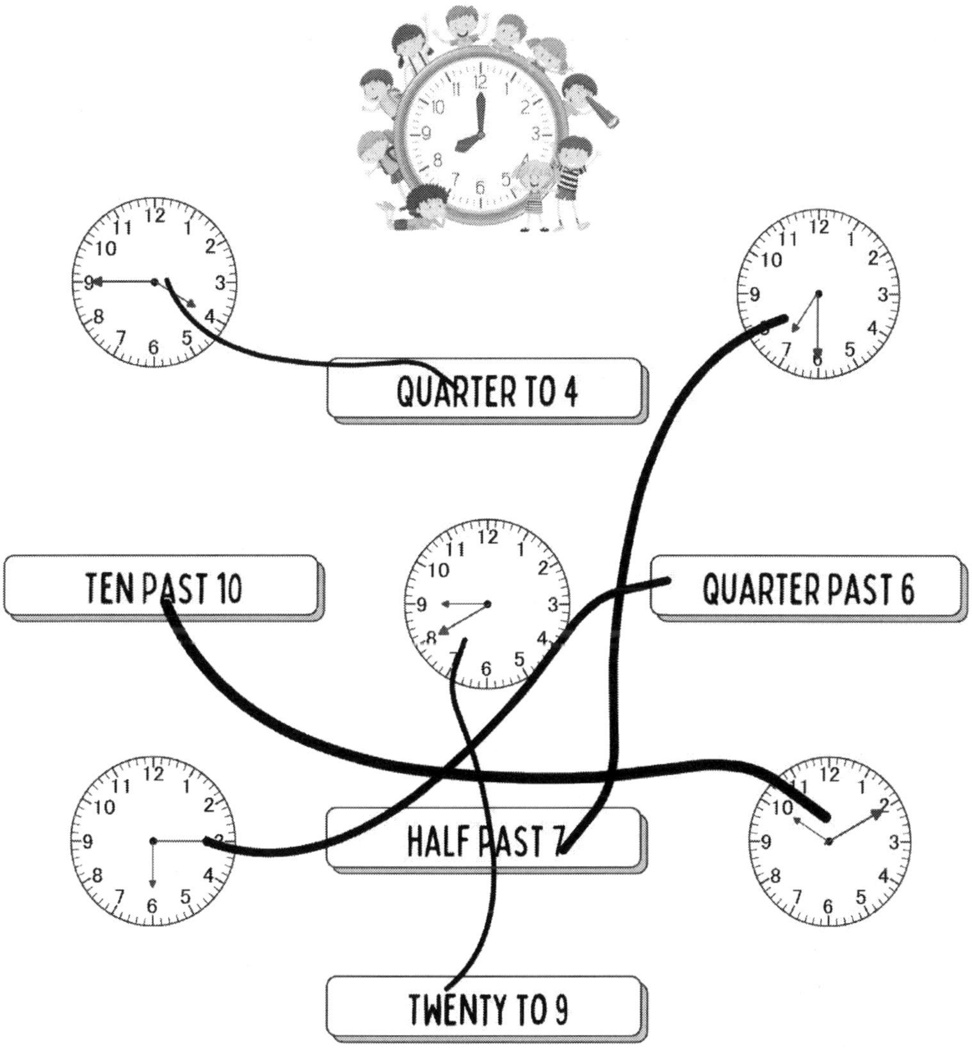

# ANSWERS TO P.39:

## MATCH THE TIME

Draw a line from the time worded in English
to the appropriate clock.

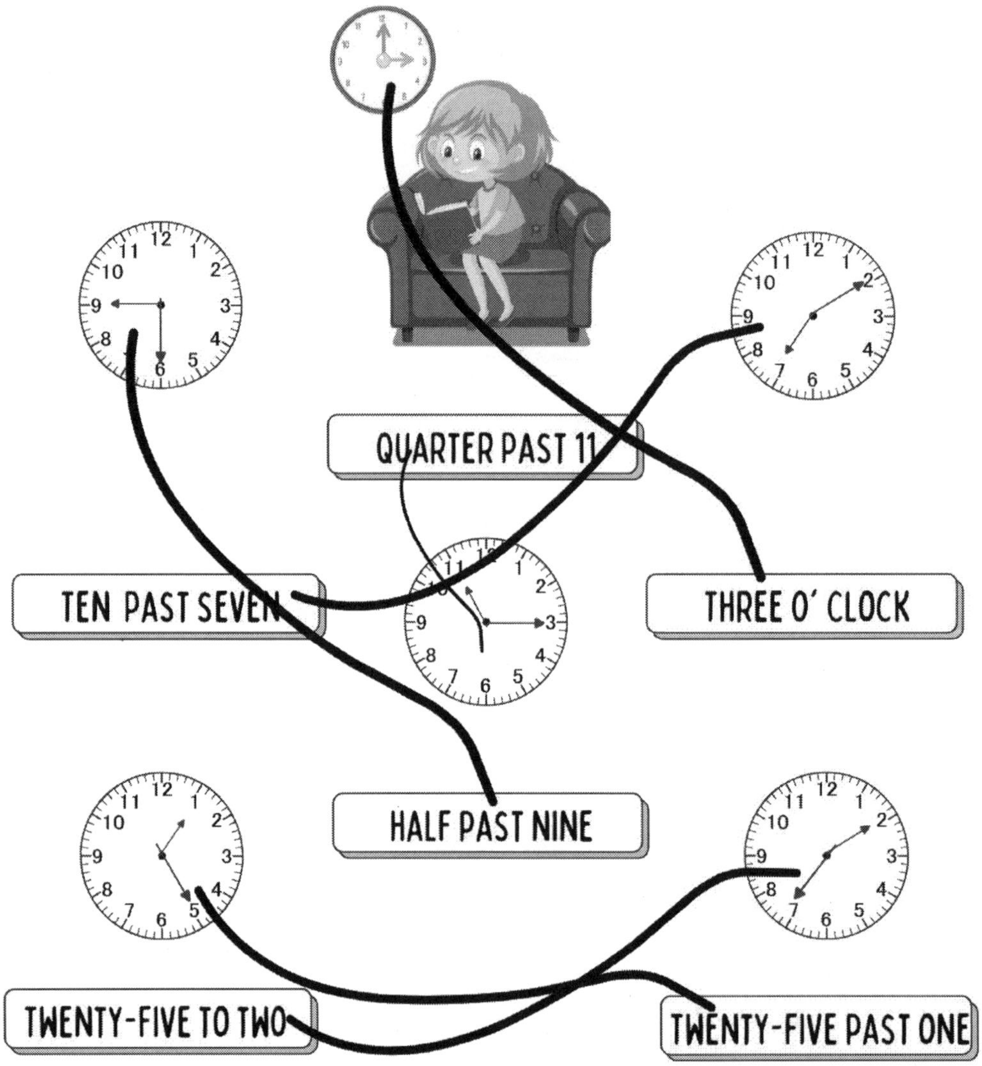

# ANSWERS TO
# MATCH THE TIME P.40-41

1. **D**
2. **C**
3. **B**
4. **A**
5. **H**
6. **I**
7. **E**
8. **J**
9. **F**
10. **H**
11. **G**
12. **D**
13. **C**
14. **B**
15. **A**
16. **K**
17. **I**
18. **E**
19. **J**
20. **F**
21. **H**
22. **G**

# ANSWERS TO P.42

## ELAPSED TIME EXERCICES

| The Present Time | 5 Minutes Before | 10 Minutes After |
|---|---|---|
| 2:50 | 2:45 | 3:00 |
| 4:00 | 3:55 | 4:10 |
| 5:15 | 5:10 | 5:25 |
| 1:30 | 1:25 | 1:40 |
| 2:15 | 2:10 | 2:25 |
| 5:45 | 5:40 | 5:55 |
| 6:10 | 6:05 | 6:20 |
| 7:50 | 7:45 | 8:00 |
| 7:20 | 7:15 | 7:30 |
| 6:35 | 6:30 | 6:45 |

# ANSWERS TO P.43

## ELAPSED TIME EXERCICES

| The Present Time | 5 Minutes Before | 10 Minutes After |
|---|---|---|
| 3:40 | 3:35 | 3:50 |
| 2:00 | 1:55 | 2:10 |
| 9:15 | 9:10 | 9:25 |
| 12:35 | 12:30 | 12:45 |
| 10:15 | 10:10 | 10:25 |
| 3:10 | 3:05 | 3:20 |
| 5:25 | 5:20 | 5:35 |
| 11:00 | 10:55 | 11:10 |
| 1:20 | 1:15 | 1:30 |
| 8:30 | 8:25 | 8:40 |

# ANSWERS TO QUIZ TIME (P.48-49)

Answers: 1) B, 2) C, 3) C, 4) C, 5) C, 6) A, 7) A, 8) A, 9) B 10) D

# ANSWERS TO WORD PROBLEMS
# CALENDER DAYS & MONTHS (P.50)

1.To find the day Sophie returns to school, we need to count 10 days starting from Monday. Counting from Monday, we have:
Tuesday, Wednesday, Thursday, Friday, Saturday, Sunday, Monday, Tuesday, Wednesday, Thursday. So Sophie returns to school on a **Wednesday.**

2.To find how many days it will take Pablo to save enough money, we need to divide the cost of the toy ($15) by the amount he saves per day ($2).
This gives us: 15 / 2 = 7.5. Since Pablo can't save half a day, we round up to the next whole number, which is 8. So it will take Pablo **8 days** to save enough money to buy the toy.

**If you do not know how to divide yet, this is the solution:**

If Pablo saves $2 each day, we can count how many days it takes for him to save up $16.

1st day: $2 saved
2nd day: $4 saved ($2 from the 1st day + $2 from the 2nd day)
3rd day: $6 saved ($4 from the 2nd day + $2 from the 3rd day)
4th day: $8 saved ($6 from the 3rd day + $2 from the 4th day)
5th day: $10 saved ($8 from the 4th day + $2 from the 5th day)
6th day: $12 saved ($10 from the 5th day + $2 from the 6th day)
7th day: $14 saved ($12 from the 6th day + $2 from the 7th day)
8th day: $16 saved ($14 from the 7th day + $2 from the 8th day)

Therefore, it will take Pablo 8 days to save enough money to buy the toy.

# ANSWERS TO WORD PROBLEMS
# CALENDER DAYS & MONTHS (P.50-51)

3. Count the number of days and multiply it by the time she wants to practice each day.

For example, Stacey has 9 days until her performance, and she wants to practice for at least 30 minutes each day.

So, she will practice for a total of:
9 days x 30 minutes/day = 270 minutes

To convert minutes to hours, the child can divide by 60, since there are 60 minutes in an hour:
270 minutes / 60 = 4.5 hours

Therefore, Stacey will practice for a total of 4.5 hours before her dance performance.

4.To find the number of days of summer vacation, we need to count the number of days between June 6th and September 2nd, inclusive. We can do this by counting the number of days from June 6th to July 6th (30 days), from July 7th to August 6th (31 days), and from August 7th to September 2nd (27 days). Adding these up, we get: 30 + 31 + 27 = 88. However, we need to include September 2nd as part of the summer vacation, so we add 1 to get 89.

Therefore, students get 89 days of summer vacation.

# ANSWERS TO WORD PROBLEMS CALENDER DAYS & MONTHS (P.51)

5. John has been saving money for 6 months. If he started saving in January, in which month will he have saved for a whole year?

To solve this problem, we need to count forward from January to find out which month is 12 months ahead. Starting with January, we have February, March, April, May, June, July, August, September, October, November, and December.

Therefore, he will have saved for a whole year in **December.**

6. If Mary's birthday is in 3 months and her birthday is in June, we can count 3 months back from June to determine the current month.

June is the sixth month of the year, so counting three months back from June gives us March. Therefore, the current month is **March**.

# ANSWERS TO ADVANCED QUIZ KIDS GRADE 3-4 (P.52)

1.To find the total amount of time Johnny spends doing his chores every day, we need to add up the time it takes him to make his bed, clean his room, and water the plants:

15 minutes (to make his bed) + 25 minutes (to clean his room) + 5 minutes (to water the plants) = **<u>45 minutes</u>**

# ANSWERS TO ADVANCED QUIZ KIDS
# GRADE 3-4
# (P.52)

2.To find the solution, we need to add the amount of time Sarah spends watching TV to the time she finished her homework:
4:30 pm + 45 minutes = 5:15 pm

Therefore, it will be 5:15 pm when Sarah is done watching TV.
The answer is **(C) 5:15 pm.**

3. We know that Katie can run 10 meters in 2 seconds. To find out how many seconds it will take her to run 50 meters, we can use repeated addition.

Since 10 meters is covered in 2 seconds, we can determine how many sets of 10 meters are needed to reach 50 meters.

10 meters + 10 meters + 10 meters + 10 meters + 10 meters = 50 meters
Similarly, we add the corresponding time:

2 seconds + 2 seconds + 2 seconds + 2 seconds + 2 seconds = 10 seconds

Therefore, it will take **Katie 10 seconds to run 50 meters** and reach the slide.

# ANSWERS TO ADVANCED QUIZ KIDS
# GRADE 3-4 (P.53-54)

4. To find the solution, we can add up the gaming times for each day of the week. Let's calculate it step by step:

Monday: 20 minutes Tuesday: 30 minutes Wednesday: 0 minutes Thursday: 35 minutes Friday: 25 minutes Saturday: 25 minutes Sunday: 0 minutes

To find the total gaming time for the week, we can add these amounts together:
20 + 30 + 0 + 35 + 25 + 25 + 0 = 135

Therefore, Joshua spends a total of **135 minutes** gaming in a week.

Another way to write it: 60 minutes + 60 minutes= 120 minutes (2 hours)
135-120 minutes=15 minutes.

Final answer **2 hours and 15 minutes**. Both answers are good!

5. To determine how long it will take Cedric to complete his chores, we need to add up the times for each task. Let's calculate it step by step:

Chickens: 5 minutes to lay their eggs. Cows: 15 minutes to be milked. Pigs: 20 minutes to eat their food. Horses: 30 minutes to be groomed.

To find the total time it will take Cedric to complete his chores, we can add these amounts together:

5 minutes + 15 minutes + 20 minutes + 30 minutes = **70 minutes**

Therefore, it will take Cedric 70 minutes to complete his chores on the farm.

# Our other books on Amazon:

https://linktr.ee/lightofmindhomeschool

# Fall Collection

## Primary Composition Notebook for Kids 5-8

SCAN ME

# CERTIFICATE
# OF
# ACHIEVEMENT

is presented to

________________________________________

**for completing this "Learning Time Workbook" with success!**

**Congratulations! Tap yourself on the back!**

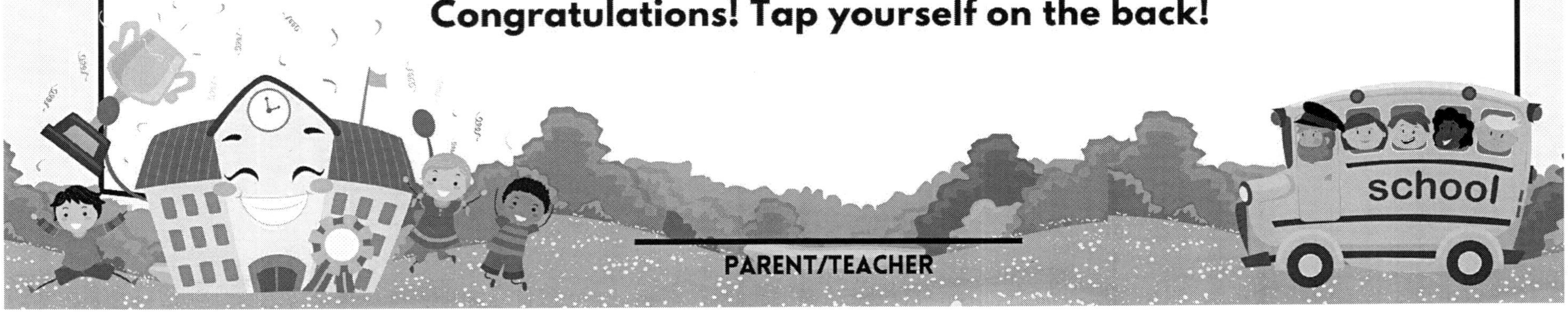


____________________

**PARENT/TEACHER**

Made in United States
Troutdale, OR
11/03/2026

56844670R00044